"A Concise Primer on Internet Radio, Webcasting and Podcasting"

Visit us at
StartupGuideTo.com

Startup Guide TO Internet Broadcasting

Brian Cochran, D.M.

A Reference For The Beginner

Startup Guide TO

Internet Broadcasting
1st Edition
A Reference for beginners!

By Brian Cochran, D.M.

Books and Manuals

An International Data Group Company

Pomona, Ca.

Startup Guide to Internet Broadcasting
Published by Biz Help 101
1107 E. Grand Ave.,
Pomona, Ca. 91766
www.StartUpGuideTo.com (website)

Table of Content

Startup Guide to
Internet Broadcasting

How Streaming Video and Audio Work

The Internet has given the common person the ability to go global, make millions and reach their target customer in a cost effective manner. Broadcast, Cable and Satellite in most cases are too expensive to use as a medium and unless you have the ability to syndicate or use satellite it's regional or local.

In the early days of streaming media -- the mid-to-late 1990s -- watching videos and listening to music online wasn't always fun. It was a little like driving in stop-and-go traffic during a heavy rain. If you had a slow computer or a dial-up Internet connection, you could spend more time staring at the word "buffering" on a status bar than watching videos or listening to songs. On top of that, everything was choppy, pixilated and hard to see.

Streaming video and audio have come a long way since then. According to Bridge Ratings, 57 million people listen to Internet radio every week. In 2006, people watched more than a million streaming videos a day on YouTube [source: Reuters]. The same year, television network ABC started streaming its most popular TV shows over the Web. People who missed an episode of shows like "Lost" or "Grey's Anatomy" could catch up on the entire thing online -- legally and for free.

The success of streaming media is pretty recent, but the idea behind it has been around as long as people have. When someone talks to you, information travels toward you in the form of a sound wave. Your ears and brain decode this information, allowing you to understand it. This is also what happens when you watch TV or listen to the radio. Information travels to an electronic device in the form of a cable signal, a satellite signal or radio waves. The device decodes and displays the signal.

In streaming video and audio, the traveling information is a stream of data from a server. The decoder is a stand-alone player or a **plugin** that works as part of a Web browser. The server, information stream and decoder work together to let people watch live or prerecorded broadcasts.

In this article, we'll explore what it takes to create this stream of ones and zeros as well as how it differs from the data in a typical download. We'll also take a look at how to make good streaming media files.

Why live streaming video?

With the invention of the Internet you can now cost effectively reach your target market. From sales presentations to sitcoms you can use your imagination and ability to market and you are able to be just like broadcast TV. I said back in 1992 when I first discovered the Internet that the Internet will be the regular TV set and will be in every car. It has come to pass and now most of the world can tune into your broadcast with a smartphone.

Churches- There is times when members of your congregation are unable to physically attend worship services. With live streaming, they don't have to miss a thing and can FEEL connected in ways that a simple audio feed cannot achieve. Now many churches are providing what is called "e-Church" or Internet Churches where they are offering Sunday service online for those to tune in anywhere online in the globe. There are Churches that we have launched where pastors where their congregation is online only!

Startups- The whole reason that I wrote this guide is because of how easy it is for a person can become a star and make a living doing what they love. Going live from your smartphone or a studio you can broadcast your message, show to your fans and pick up new ones. Not to many years ago, you had to have very high tech equipment and a big budget in order to broadcast live, but now as stated I know of shows broadcasting live with their iPad and smartphone.

Live Events- From Football to MMA you can live stream your event and do it with very little equipment. You can set up paid per view event that you can live stream the event and generate extra income. Live news, entertainment shows to red carpet events just like broadcast TV you can produce the same type of quality streaming on the Internet.

I've event did a live wedding and funeral for family and friends that could not attend so it was broadcasted live for others to view. The Internet has made it possible for the impossible.

Why Internet Radio?

Just like Streaming Video, Radio is a powerful means to get you content to the masses. I have used both traditional and Internet and they both have their pros and cons. The pros for Internet radio is the cost and the global reach. The biggest con is that you can yet charge for commercials like traditional radio. Traditional Radio is costly to start, run and you can't even start a station on you own, you have to purchase a pre-existing station and it will cost millions.
From the comfort of your bedroom you can launch a station that has thousands to millions of listeners with a small to none budget. A laptop and DSL or broadband and you one. There are so many online streaming companies that you can launch a show in minutes.
Internet Radio still has it's place for instance in the workplace where a person can listen while their work. Millions listen in their car, bus, train from their smartphone. Technology has revolutionized they way people are getting information and entertainment. As mentioned earlier, The Smartphone and iPad/Tablet has made where a person that has a broadcast can reach more people than ever and don't have to depend on traditional means to reach the public.
As a small business and you have a product or service that you want to get to market, Internet radio is a wide open field and is one of the most cost effective ways to reach your target market. You are finding that even AM and FM stations are using the Internet to broadcast their stations. Even offering Internet Only advertising to local businesses.

We will go into greater details about both Video and Radio Broadcasting; I just wanted you to get a little heads start on them. Hopefully by the end of the guide you will start developing your broadcast.

Finding and Playing Streaming Video and Audio

If you have a connection to the Internet and you want to find streaming video and audio files, you shouldn't have to look far. Sound and video have become a common part of sites all over the Web, and the process of using these files is pretty intuitive. You find something you want to watch or hear -- you click it, and it plays. Unless you're watching a live feed or a **webcast**, you can often pause, back up and move forward through the file, just like you could if you were watching a DVD or listening to a CD.

But if you've never used streaming media, your computer may need a little help to decode and play the file. You'll need a **plugin** for your Web browser or a **stand-alone player**. Most of the time, the Web page you've visited points you in the right direction. It prompts you to download a specific player or shows you a list of choices.

These players decode and display data, and they usually retrieve information a little faster than they play it. This extra information stays in a **buffer** in case the stream falls behind. There are four primary players, and each one supports specific streaming file formats:

- QuickTime, from Apple, plays files that end in .mov.
- RealNetworks RealMedia plays .rm files.
- Microsoft Windows Media can play a few streaming file types: Windows Media Audio (.wma), Windows Media Video (.wmv) and Advanced Streaming Format (.asf).
- The Adobe Flash player plays .flv files. It can also play .swf animation files.

For the most part, these players can't decode one another's file formats. For this reason, some sites use lots of different file types. These sites will ask you to choose your preferred player or pick one for you automatically.

The QuickTime, RealMedia and Windows Media players can work as stand-alone players with their own menu bars and controls. They can also work as browser plugins, which are like miniature versions of the full-scale player. In plugin mode, these players can look like an integrated part of a Web page or pop-up window.

Flash video is a little different. It usually requires a Flash **applet**, which is a program designed to decode and play streaming Flash files. Programmers can write their own Flash applets and customize them to fit the needs of a specific Web page. Flash is becoming a more popular option for playing streaming video. It's what YouTube, Google Video and the New York Times all use to display videos on their sites. The video below, which demonstrates what would happen if you shot your TV, plays in a Flash applet.

Regardless of whether it's an applet or a fully functional player, the program playing the streaming file discards the data as you watch. A full copy of the file never exists on your computer, so you can't save it for later. This is different from **progressive downloads**, which download part of a file to your computer, then allow you to view the rest as the download finishes. Because it looks so much like streaming media, progressive downloading is also known as **pseudo-streaming**.

These players and applets do what many applications do -- they play files. We'll look at these files and how they travel to your computer in the next section.

Streaming Files

Streaming video and audio files are compact and efficient, but the best ones start out as very large, high-quality files often known as **raw files**. These are high-quality digital files or analog recordings that have been digitized, and they haven't been compressed or distorted in any way. Although you can watch a streaming file on an ordinary TV, editing the raw file requires lots of storage space and processing power.

It might seem strange that a file that ends up nimble and efficient started out large and unwieldy. The reason is that the compression process, required to make an ordinary file into a streaming file, lowers the file's quality. During compression, blurry, low-quality videos or hard-to-hear audio recordings will only get worse.

Fortunately, before you even compress a file, you can reduce its size without lowering its quality:

- **Make the picture smaller:** Most streaming videos don't fill the whole screen on a computer. Instead, they play in a smaller frame or window. If you stretch many streaming videos to fill your screen, you'll see a drop in quality.

- **Reduce the frame rate:** A video is really a series of still images. The frame rate is how quickly these images move from one to the next. A lower frame rate means fewer total images and less data needed to recreate them. The reduction in frame rate is why some streaming videos flicker -- the frame rate is slow enough that your eye and brain sense the transitions between pictures.

For both video and audio files, making the files even smaller requires **codec,** or **compression/decompression** software. Codecs discard unnecessary data, lower the overall **resolution** and take other steps to make the file smaller. Different codecs also create specific types of files, which work on specific streaming players.

The total reduction in quality depends on a number of factors, including the **bitrate,** or the speed of the transfer from the server to a computer. For example, the bitrate of a television broadcast is about 240,000 kilobits per second (Kbps), but the bitrate of a dial-up Internet connection is a maximum of 56 Kbps. Someone with a reliable broadband connection with lots of bandwidth can watch high-bitrate files, but someone using a dial-up modem needs to watch at a much lower bitrate. The basic idea is to encode a file that's large enough to look or sound good but small enough to work with the available bandwidth. Some codecs let you create files that will stream differently at different transfer rates, accommodating different connection types. This is known as **multi-bitrate encoding.**

Once a file is edited, compressed and encoded, it's uploaded to a server. We'll look at the server's role in streaming media in the next section.

CREATING GOOD STREAMING VIDEOS

Keep it simple -- the more complexity you put into your shot, the more detail the computer will have to render later.

* Use a steady, unobtrusive background. If you have a green screen, use it -- you can add a different background during editing.
* Keep the camera still.
* If you're filming people, make sure they wear solid colors rather than patterns.

Streaming Servers

If you work in an office that shares files over a network, you might think of a server as a computer that holds lots of data. But when it comes to streaming video and audio, a server is more than just a massive hard drive. It's also the **software** that delivers data to your computer. Some streaming servers can handle multiple file types, but others work only with specific formats. For example, Apple QuickTime Streaming Server can stream QuickTime files but not Windows Media files.

Streaming servers typically deliver files to you with a little help from a Web server. First, you go to a Web page, which is stored on the Web server. When you click the file you want to use, the Web server sends a message to the streaming server, telling it which file you want. The streaming server sends the file directly to you, bypassing the Web server.

All of this data gets to where it needs to go because of sets of rules known as **protocols**, which govern the way data travels from one device to another. You've probably heard of one protocol -- hypertext **transfer protocol (HTTP)** deals with **hypertext** documents, or Web pages. Every time you surf the Web, you're using HTTP.

Many protocols, such as **transmission control protocol (TCP)** and **file transfer protocol (FTP)**, break data into **packets**. These protocols can re-send lost or damaged packets, and they allow randomly ordered packets to be reassembled later. This is convenient for downloading files and surfing the Web -- if Web traffic slows down or some of your packets disappear, you'll still get your file. But these protocols won't work as well for streaming media. With streaming media, data needs to arrive quickly and with all the pieces in the right order.

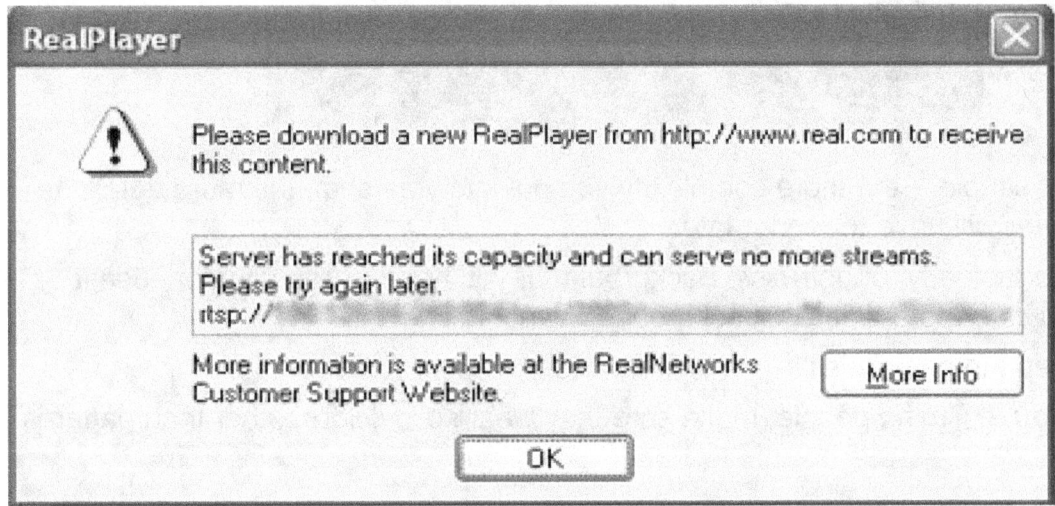

Too many outgoing streams can overload a server, causing users to see an error message.

For this reason, streaming video and audio use protocols that allow the transfer of data in **real time**. They break files into very small pieces and send them to a specific location in a specific order. These protocols include:

- Real-time transfer protocol (RTP)
- Real-time streaming protocol (RTSP)
- Real-time transport control protocol (RTCP)

These protocols act like an added **layer** to the protocols that govern Web traffic. So when the real-time protocols are streaming the data where it needs to go, the other Web protocols are still working in the background. These protocols also work together to balance the load on the server. If too many people try to access a file at the same time, the server can delay the start of some streams until others have finished.

STREAMING CHOICES

- **Live or on-demand:** Live webcasts require some extra equipment. You'll need an on-site computer that can compress, encode and stream the video feed in real time or a satellite uplink to a company that can do it for you.
- **Unicast or multicast:** In a unicast stream, each person watching gets his own stream of data. In a multicast stream, one stream of data travels to a router, which copies the stream and sends it to multiple viewers. Unicast streams require more processing power and bandwidth.

Step-by-step Streaming

Using streaming media files is as easy as browsing the Web, but there's a lot that goes on behind the scenes to make the process possible:

1. Using your Web browser, you find a site that features streaming video or audio.
2. You find the file you want to access, and you click the image, link or embedded player with your mouse.
3. The Web server hosting the Web page requests the file from the streaming server.
4. The software on the streaming server breaks the file into pieces and sends them to your computer using real-time protocols.
5. The browser plugin, standalone player or Flash application on your computer decodes and displays the data as it arrives.
6. Your computer discards the data.

All of this requires three basic components -- a player, a server and a stream of data that are all compatible with each other.

Creating and distributing a streaming video or audio file requires its own process:

1. You record a high-quality video or audio file using film or a digital recorder.
2. You **digitize** this data by importing it to your computer and, if necessary, converting it with editing software.
3. If you're creating a streaming video, you make the image size smaller and reduce the frame rate.
4. A codec on your computer compresses the file and encodes it to the right format.
5. You upload the file to a server
6. The server streams the file to users' computers.

Because of advances in home computers and software, it's become easier for people to create their own streaming videos at home. Most people can't afford to purchase and maintain their own streaming servers and instead pay a **service provider** to host the videos. But the increased availability of streaming video has also created some challenges. One is **copyright.** It's easier than ever to illegally copy TV shows or other videos and post them on the Web, and legal action from copyright owners has become more common.

Another challenge has to do with **royalties**. Streaming video has changed the way people watch TV shows and movies, and some actors, writers and other entertainment industry workers claim they aren't being paid as they would for TV broadcasts or theater screenings. In addition, in March 2007, the U.S. Copyright Royalty Board changed its royalty structure, making Internet radio far more expensive to produce than it had been.

In spite of these complications, the world of streaming video and audio continues to grow. In the next few years, Internet TV, Internet radio and other streaming applications may become real competitors against traditional media.

If you'd like to learn more about streaming video, streaming audio and related topics, you'll find lots of resources on the next page.

Streaming video is content sent in compressed form over the Internet and displayed by the viewer in real time. With **streaming video** or **streaming** media, a Web user does not have to wait to download a file to play it. Instead, the media is sent in a continuous **stream** of data and is played as it arrives.

5 Ways To Use Video To Create A Connection With Your Audience Online

Think for a moment about all the websites that you visit when you're on the web. Do the majority of them have some sort of video incorporated into their website?

There are constantly new ways to incorporate video into your website. If you're not using video yet in some way, maybe it's time to jump on the bandwagon because it is surely not going away.

1. Interviews
   a. One of the easiest ways that you can produce content on the web is to conduct an interview. An interview is where you ask questions on video with a person(s) about a given subject.

2. Testimonials
   a. Testimonials are where customers have used your products or services and they share what they got out of using them. Very powerful, it's just like using word of mouth, people support what their peers have used and liked.

3. Video tutorials
   a. Now this is somewhat similar to the how-to videos, but video tutorials are more for explaining and demonstrating the product that you are offering.

4. An introduction to your company
    a. It's important that there is a face to your ministry
    b. One of the best ways to convey you being the face and leader of your business is to do a welcome video on the home page of your website.

5. Live Video Broadcasting
    a. I see live video streaming is pretty popular, for the simple fact that you can have a huge amount of interaction with your viewers.
    b. I love it when my favorite experts take the time to do some simple Q&A while they have a free moment.

Broadcasting 101

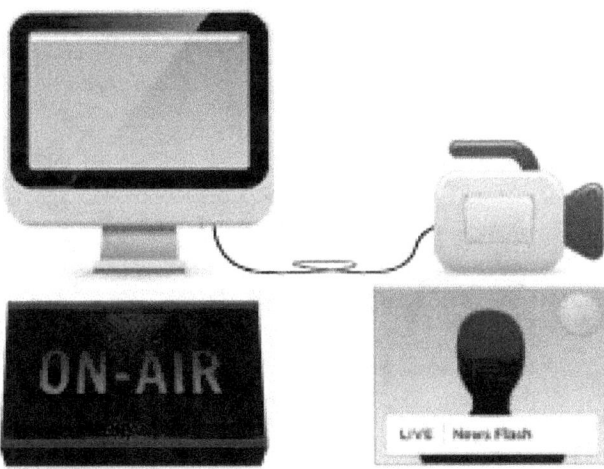

In order to produce a show you will need some equipment and in some cases software will be needed as well. Just to let you know, this is a broad stroke and depending on your plan for your broadcast is what you will need. Oh yeah, also budget as well.

Tools needed for broadcasting a show or station:
- Video
  - Camcorder
  - Webcam
  - Camera with video feature
  - Smartphone
  - iPad/tablet
- Computer system
  - Laptop
  - PC computer
- Audio
  - Microphone
  - Headset
  - Monitor system

Live Video Broadcast Production

Video productions using multiple cameras, pre-recorded and live video and audio, video effects, overlays, Chroma Keyers and more, while simultaneously recording your production and streaming live on the Internet.

- **VidBlaster (**Windows**)**
  VidBlaster is the ideal affordable live video production software offering broadcast resolution up to HD with stunning video quality. Easy to use, Vidblaster is modular based, allowing you to customize your productions for any live event.

- **StudioCoast VMix HD Pro** (Windows)
  vMix is a new Software Video Mixer and Switcher that utilizes the latest advances in computer hardware to provide live HD video mixing, a task previously only possible on expensive dedicated hardware mixers. vMix runs on the Windows XP, Windows Vista and Windows 7 platforms.

- **CamTwist (**Mac**)**
  CamTwist is a software package that lets you add special effects to your video chats. It's also possible to stream your desktop and still images. With CamTwist, you can also use multiple video chat programs at the same time.

CamTwist works with Stickam, justin.tv, ustream.tv, operator11, yahoo, skype and amsn. Ustream.tv, Justin.tv, BlogTV, LiveStream, NuMuBu.com and many others you can stream your CamTwist produced shows live on the Internet. CamTwist does NOT work with iChat. Want higher quality? CamTwist fully supports Flash Media Live Encoder as well as Telestream's Wirecast allowing you to stream HD quality content live online. Don't want to download and install software? No problem. CamTwist can be seen by most modern browsers as an integrated webcam so you can select it as a source in the Adobe Flash Player.

- **ManyCam** (Windows and Mac)
ManyCam Pro turns ManyCam into a Pro video switcher with transitions and more! HD Video Support and Broadcasting. Quickly take photo snapshots with one click. Ability to capture and display specific applications. Add Lower Thirds to your video screen. Use your webcam with many
applications simultaneously, Such as use Skype, MSN, Ustream, and many other webcam applications at the same time.

ManyCam allows you to- Use your webcam with multiple programs simultaneously. Add text to your webcam video window with any application. Add cool animations to your video window. Show your local day and date in your video window. Add live CGI graphics like fire and water effects, make it appear as if it is snowing inside your house and more.

Recording and Pre-recording your show or broadcast

Depending on what you are using for broadcasting it may not have recording within the system or service. The following are some of the products that allow you to record your screen, camera and desktop.

Live Streaming Recording

- **Debut Video Capture Software**
  Record video from a webcam, recording device or screen. Capture video files on your Mac or PC with this easy video recorder software. Great low cost software that does a great job.

- **Screen Recorder Studio**
  Get your video on the web with Video Flow software. With Video Flow you can record the contents of your entire monitor while also capturing your video camera, microphone and your computer audio. The easy-to-use editing interface lets you creatively edit your video; add additional images, text, or music; and add transitions for a truly professional-looking video. The finished result is a QuickTime Media movie, ready for publishing to your Web site, save to the local or work seamlessly with iDVD to burn your video to DVD.

- **VideoPad Video Editor**
  VideoPad Video Editor is a Mac Video Editing Software Anyone Can Use Designed to be intuitive, VideoPad is a fully featured video editor for creating professional quality videos in minutes. Making movies has never been easier. Edit video from any camcorder
  Capture video from a DV camcorder, VHS, webcam, or import most any video file format including avi, wmv, mpv and divx. Full of transitions and visual effects. Over 50 visual and transition effects to add a professional touch to your movies. Create videos for DVD, HD, YouTube and more Burn movies to DVD for playback on TV, or as a standalone video file to share online or put on portable devices.

- **Adobe Premiere** (Both Mac and Windows)
  Adobe Premiere is the Film and Broadcast industry's primary editing and production program for film and video, etc. Used in conjunction with Adobe After Effects (which is used to add titles, transitions, effects, etc.), it's used to create most of the multimedia you see on a daily basis.

- **Sony Vegas Pro 13** (Windows)
  Sony Vegas Pro 13 Edit is an outstanding professional video editing application. It contains all the same editing tools found in the so-called industry-standard applications. It's a great alternative to the big names, especially if you capture your footage with Sony cameras. And aside from some minor cosmetic issues, there isn't much to complain about.

- **Wondershare Video Editor** (Mac & Windows)
  Wondershare Video Editor is the most user-friendly home video editing software, featuring smart and intuitive editing tools that let you create Hollywood-themed movies within minutes.
  Cut, edit, merge, and trim clips. Add music and text. Apply special effects. Get a professional-looking movie in minutes.

- And many more…

There may be shows that you would like to pre-record as well and there are several programs that do it in a cost effective way.

Screen Capture and more

- **Photo Booth** for Mac & Windows
Photo Booth is the non-official Windows port of the popular Mac app for taking photos with your webcam.   Photo Booth application, which allows you to add fun and silly effects to your photos and videos.

   Like the original app, Photo Booth for Windows includes a selection of special effects you can apply to the image on your webcam in real time, and then take a picture of it. The good thing about Photo Booth for Windows 7 is that it doesn't require installation and is very easy to set up. All you need is Adobe Flash Player and a webcam. Then you can start applying effects and taking snapshots of yourself and your friends posing for the webcam.

   Though Photo Booth for Windows lacks some features found in the original program – such as background effects, or being able to record video – it's still one of the best Photo Booth clones for Windows you can find these days. Photo Booth is an easy, addictive app with which to have fun with your webcam.

- **Camtasia** (Mac & Windows)
Camtasia is powerful, yet easy-to-use screen recorder, Camtasia helps you create professional videos without having to be a video pro. Easily record your screen movements and actions, or import HD video from a camera or other source. Customize and edit content on both Mac and Windows platforms, and share your videos with viewers on nearly any device.

- **VodBurner** (Mac & Windows)

VodBurner is used to record Skype Video & Audio. Edit your videos in split-screen (side by side), or this side / other side camera views with our Post-Production Console. Add subtitles and more.

- **ScreenFlow** (Mac)
  ScreenFlow is a very polished screencasting application, which records just about anything and gives you a huge amount of editing options afterwards. Screencasts, which are videos of your computer screen and often used in tutorials, are a useful way to illustrate a point or program. ScreenFlow will record everything from your iSight camera to microphone audio or speaker audio enabling you to add crystal clear commentaries to your screencasts. ScreenFlow even monitors keystrokes and mouse movements, which is pretty clever. The app also has an excellent zoom function which allows you to add a touch of class and professionalism to your screencasts. Recording quality is excellent, and thanks to a high frame-rate (which you can set), your videos look as if everything was actually happening on your screen. What's also surprising is that ScreenFlow is very lightweight for what is effectively a video editor, consuming relatively few CPU and RAM resources. Exporting is possible in all major formats and YouTube fans can now share their productions instantly. The program has one glaring drawback however and that's that you can't add text, subtitles or simple text annotations to your beautifully produced screencasts. It's also a bit on the pricey side but if you're looking to produce professional looking screencasts, it's probably worth it. ScreenFlow is an excellent screencasting application that just lacks text insertion support to make it a truly excellent application.

- **Screen Recorder Studio** (Mac)
  Screen Recorder Studio is a professional screen record software, what you see is what you get. The app can record computer audio and you can record online video. It is easy to record your screen, which capture computer audio, record audio from built-in such as Mic .
  Screen Recorder Studio can record single window, custom area or entire screen. You can demo your software or record online video .
  Screen Recorder Studio supply great text logo and image logo. You can make beautiful logo. You can add camera on screen and add the mouse click animation.
  Screen Recorder Studio can custom desktop background and hide desktop icons when you are recording, Screen Recorder Studio provides all the elements needs to create professional-looking product demonstrations of software applications, can captured games screen, PPT, your image's slideshow etc…

Post Production Software

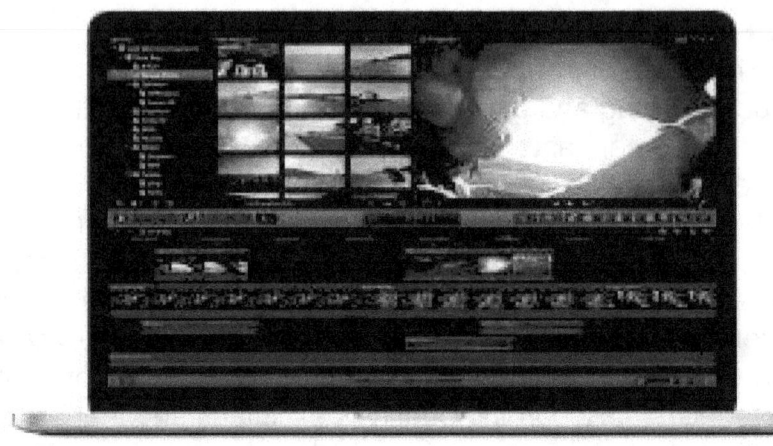

One you have recorded your raw video footage you will in most cases need to do some editing and the following software are some of the most popular editing/post production resources you can use.

TIP

Post production software is really key to use when you want to add Intros which is introducing your video, host, subject and so on, and outros which are the credits, call to actions and so on…

- **Final Cut Pro X** (Mac)
  Final Cut Pro offers an exceptionally fluid, flexible way to edit. Assemble shots with ease as clips "magnetically" close up to eliminate unwanted black gaps in the timeline. Similarly, clips move out of the way to avoid clip collisions and sync problems. If you prefer to work with traditional, non-rippling editing behaviors, you can use the Position tool to build your timeline. Interactive animations show you exactly what's happening as you work, so you can easily try out new ideas.

- **iMovie** (Mac)
  iMovie is a software that allows users to create a digital movie. Footage is incorporated from a video camera, DVDs or VHS, recorded sound, and still images.

- **Video Flow** (Mac)
  You can get online video or online music with Video Flow software. Fire up Video Flow, webpage, or Keynote presentation that you want to appear in your screencast video, and the screen recorder captures everything that happens on your screen while also capturing your video camera, microphone and your computer audio. Video Flow can record screen by three format (H264 ,Mp4, Animation) ,the three formats have their own advantages, you can set by your request. Choose a custom recording area or full screen record. Use Video Flow to create high-quality software demos, tutorials, app demos, training, presentations and more!
  Video Flow is a video editing software can help you to edit, merge and retouch videos. The editing functions contain trimming, Splitting, adding stylish subtitle, applying filter effects, inserting transition etc. You can voiceover for the video.

- **Avid**
  Avid's industry-leading video editing solutions provide the cutting-edge tools and streamlined workflows you need to maximize your creativity and tell the most compelling story possible. Featuring software and hardware that integrate seamlessly, Avid video editing solutions speed up your workflow and free you from technical distraction.

- **Movie Maker** (Windows)
  Movie Maker, it's easier than ever to create and share your own movies. It's part of Windows Essentials, a free download that also includes tools for photos, instant messaging, email, social networking, and more.

Video Encoder and Encoding

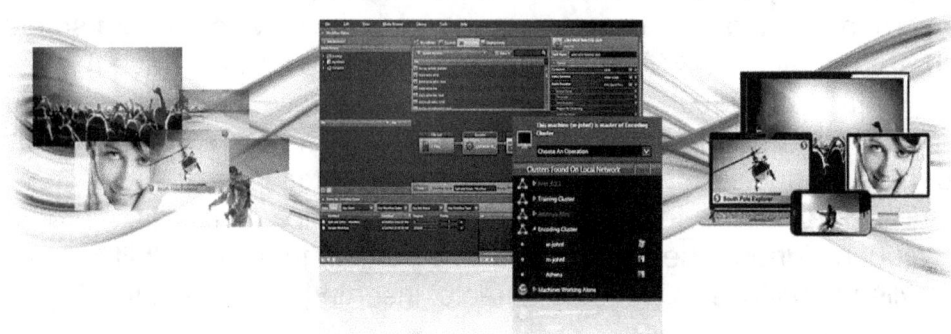

In video editing and production video encoding is the process of preparing the video for output, where the digital video is encoded to meet proper formats and specifications for recording and playback through the use of video encoder software. (Also called *video conversion*)

Encoding

(1) In computer technology, encoding is the process of putting a sequence of characters into a special format for transmission or storage purposes.

(2) The term used to reference to the processes of analog-to-digital conversion, and can be used in the context of any type of data such as text, images, audio, video or multimedia.

Video Storage & Sharing

You have to store your content that you have created, so that you can share or broadcast prerecorded content.   Most websites give you limited space for only sharing on your own website. The largest video storage is YouTube and it gives you the flexibility to store (host) your content but also be able to share it on other sites by links or embed code.

Video storage sites allow you to send your captured images or videos directly to Facebook, Twitter, and more. Or, quickly share your captures through email or instant message via the embed code they give you.

There are many free sites and I will give you a few to get you started:

- Vimeo
- YouTube
- Dailymotion
- Flickr
- GodTube
- Myspace
- MyVideo
- Veoh
- Viddler
- Facebook
- Metacafe
- Twitch

Video Streaming

Streaming video is content sent in compressed form over the Internet and displayed by the viewer in real time. With streaming video or streaming media, a Web user does not have to wait to download a file to play it. Instead, the media is sent in a continuous stream of data and is played as it arrives.

Sites that stream live:
- Dacast
- YouTube
- UStream
- LiveStream
- Fora
- Amazon
- Apple
- Roku
- Google Hangouts
- TeleStream
- StickAm
- BlogTV
- Justin TV
- Go To Meetings
- IM Tool Suite

Video Podcasts

Video podcasts -- also called videocasts, vidcasts and vodcasts -- combine the audio component of podcasting with visual media. This technology provides a forum for a wide variety of video podcasters, including filmmakers and artists, vloggers (video bloggers), and even those who just like sharing their videos.

Video podcasting is also helping build business, especially in the sales and marketing sectors. Through video podcasts, businesses both large and small can advertise their wares and services in a modern, cost-effective way. In the past, big businesses had better access to expensive studios where sophisticated advertisements were produced, but now even the smallest businesses can create high-quality media with just a <u>camera</u>, editing software and the Internet.

You watch a video podcast in much the same way as you would listen to a podcast.

To Watch a Video Podcast:

1. Locate a video podcast a directory or search engine, like Videocasting Station, Vodstock, and Open Media Network.
2. Download the video podcast to your portable media player or click the hyperlink.
3. Just like you do with your favorite podcasts, you can use an RSS Feed to download the latest updates of the video podcast.

Creating a video podcast naturally requires more effort than creating a podcast because you're adding the element of visual media. Therefore, video podcasting also requires some extra components, including a video camera, editing software, video encoder and video host.

To Create a Video Podcast

1. Film your video.
2. Upload your video to your computer.
3. Using editing software, add special effects and graphics or correct any problems in the video.
4. Determine whether your video will be streaming or downloadable, and use a video encoder to format it in a manageable file size for online viewing.
5. Find a host for your video podcast. Make sure that the host can accommodate your video's bandwidth.
6. If your host does not provide an RSS feed for your video podcast, create one yourself.
7. Lastly, promote your video podcast just as you would a podcast.

The podcasts and video podcasts you find online range from the amateur to the streamlined and sophisticated. They are a testament to the accessibility of this technology for listeners and creators alike. Anyone can -- and clearly anyone will -- podcast.

Vlog

The definition of VLOG or VLOGGING; a blog that contains video material.
A **vlog** (or video **blog**) is a **blog** that contains video content. The small, but growing, segment of the blogosphere devoted to **vlogs** is sometimes referred to as the vlogosphere. Some bloggers have included video content for years. Being that video has taken the Internet by storm, Vlogging has become commonplace.

Why Be a Vlogger?
With the age of video, YouTube is one the largest search engines in the world. If you are looking for how to change a light bulb lookers are going to YouTube to find it.

- Did you know that Millionaires are made from Vlogging? The Top YouTubers are started with a basic Vlog and grew into a full-on channel with millions of subscribers and views.

- More than 1 billion unique users visit YouTube each month

- Over 6 billion hours of video are watched each month on YouTube—that's almost an hour for every person on Earth

- 100 hours of video are uploaded to YouTube every minute

- 80% of YouTube traffic comes from outside the US

- YouTube is localized in 61 countries and across 61 languages

- According to Nielsen, YouTube reaches more US adults aged 18-34 than any cable network

- Millions of subscriptions happen each day. The number of people subscribing daily is up more than 3 times since last year, and the number of daily subscriptions is up more than 4 times since last year.

As you see using vlogging as a way to reach people with your message. A lot of blog sites are upgrading their sites to become video friendly so that bloggers can become up to date with the times. More people are using smartphones and iPads and are using YouTube as the main source of information and entertainment.

YouTube launched not to long ago live streaming on their site and that's a game changer. Most sites allow you to broadcast up to a small amount for free and to broadcast to a larger audience you have to pay for the service. But YouTube, is free and you can broadcast to ten's of millions for free and that can also check out your Vlogs after viewing your broadcast.

YouTube and Search Engines
Being that Google owns YouTube, they have intergraded into their search videos to almost rate higher than text. Specially YouTube content providers! When you start your YouTube Channel you will need to make sure you do the Meta tags and descriptions so that the search engines will pick up the text in your channel and videos.

 **TIP** *Closed Captions*

YouTube has what is called Closed Captions for the hearing impaired. So that the words that are being spoken will appear on the screen while the video is playing. Within YouTube you can upload the transcript of the video and it will automatically work with your video.

Now, to really rate high using Closed Captions will act like your video is an article and the spiders in the search engines will pick you your caption text and based on what is said in the video when a person is looking for your subject Google will place you higher than most of the others that don't use Closed Captions.

There are companies that you can solicit that will convert your video into text.

The following are several software and companies that you can make Closed Captions for your videos:

Software
- SynchriMedia
- Subtitle Workshop
- MovCaptioner
- And many more…

Companies
- ABERDEEN CAPTIONING, INC.
- AEGIS RAPIDTEXT, INC.
- CAPTION DEPOT, INC.
- COMPUSCRIPTS CAPTIONING, INC.
- And many more…

How to be a Vlogger

Many people like to make videos about interesting topics, their opinions and points of views or just recording their daily life. Let's take a closer look at the world of video blogging.

The largest site to host vloggers is YouTube.com and this lesson will be on how to use YouTube to get your message out.

1. **Think of a topic to make vlogs about.** It can be anything, as long as it's not boring, insulting, or illegal. Take other video bloggers as an inspiration, look throughout YouTube and you will see examples for you to learn from. But you have to be unique and true to yourself. Don't be a copycat - don't do what everyone else does.

2. **Get a YouTube account, and give your channel an attractive name.** Make your channel interesting. Make sure you think over your username, though, because it's nearly impossible to make it big with a jumbled, unreadable username with a bunch of numbers.

3. **Make about ten or more REALLY GOOD videos.** They really have to be good, and interesting. Post one or two daily for a little while, and then go daily or every other day after that. Really try not to skip two days, because what you give your viewers becomes their standard.

4. **Post one of those videos to a related, well-known video as a video response.** You will get at least a fifth of the views of the original video! Awesomeness TV on YouTube allows you to post video responses to their how to be a YouTube star series.

5. **Keep being active - nobody is going be interested if you don't post any videos for years.** If possible, try and set a specified day that you will upload videos. For a vlogger, at least three or four times a week. If this is too much, try and do one per week.

6. **Edit your videos!** This means the difference of viewers scrolling past you and clicking on your videos. Look at tutorials on using the software you have. Windows has Windows Movie Maker preinstalled, and all Apple products have iMovie available. Android has Movie Maker preinstalled.

7. **If you learn to edit, you can also set your banner, video thumbnail, and avatar.** If you're a YouTube partner, a whole new world of options opens up for your videos. Setting an avatar is easy, but if you want a banner, you'll need to pick a picture and edit it within computer margins. If you're a YouTube partner, you can also have the option to create a custom video thumbnail. Make sure that, when you edit, you have text, a close-up of your face, and a really cool background.

8. **Once you have a good reputation in the YouTube community, try to become a YouTube partner.** You will need to have at least a few thousand views to do so. YouTube gives you money for allowing ads to be displayed on your videos, and that's how great Vloggers earn money with their videos! Your videos will appear often in search results as well.

9. **Communicate with your followers.** Don't shut them out or ignore them. Take time to respond to comments, messages, and video responses you receive. This leaves a good impression and your followers will appreciate it.

Web Conference

A wide variety of web conferencing programs are available on the market. The simplest use chat and instant messaging programs host text-based group discussions. More sophisticated programs exchange visual information using webcams and streaming video and allow people to share documents online. Some programs are entirely computer- and Internet-based, and others use the telephone system to distribute audio content.

Web conferencing programs combine tools already common to web pages and Internet communication. They bundle these tools into one interface to create an interactive meeting environment. These tools include:

- HTML, XML and ASP markup
- Java scripts
- Flash animation
- Instant messaging
- Streaming audio and video

Companies can either **purchase** conferencing software and host their meetings themselves or use a **hosting service**. Hosting services provide the software and server space on which to conduct meetings. Either way, the company or the hosting service must have software to coordinate the meeting and ample server space and bandwidth to accommodate it.

Some programs can merge with a company's existing e-mail, calendar, messaging and office productivity applications. Some allow attendees to view the presentation in their regular web browser without installing any additional software. Depending on the software, people can:

- View slide presentations from programs like PowerPoint.

- Draw or write on a common **whiteboard** by using their computer mice or typing.

- Transmit still pictures or video to other attendees via a **webcam**. (This increases the required bandwidth and can sometimes slow the transfer of the presentation.)

- View information from the moderator's computer desktop using screen sharing.

- Share documents, often even if attendees don't have the software that created them, using **application sharing**.

- Ask and answer questions through audio chat (as an integrated part of the software) or by phone.

Programs include options for security and encryption. Most require moderators and attendees to use a login name and password to access the meeting. Some use SSL or TLS encryption to protect data. Some companies host web conferences on internal servers so that the data stays behind the corporate firewall.

Internet Radio

How Internet Radio Works

Internet radio has been around since the late 1990s. Traditional radio broadcasters have used the Internet to simulcast their programming. But, Internet radio is undergoing a revolution that will expand its reach from your desktop computer to access broadcasts anywhere, anytime, and expand its programming from traditional broadcasters to individuals, organizations and government.

Freedom of the Airwaves

Radio broadcasting began in the early '20s, but it wasn't until the introduction of the transistor radio in 1954 that radio became available in mobile situations. Internet radio is in much the same place. Until the 21st century, the only way to obtain radio broadcasts over the Internet was through your PC. That will soon change, as wireless connectivity will feed Internet broadcasts to car radios, PDAs and cell phones. The next generation of wireless devices will greatly expand the reach and convenience of Internet radio.

Uses and Advantages

Traditional radio station broadcasts are limited by two factors:

- the power of the station's transmitter (typically 100 miles)
- the available broadcast spectrum (you might get a couple of dozen radio stations locally)

Internet radio has **no geographic limitations**, so a broadcaster in Kuala Lumpur can be heard in Kansas on the Internet. The potential for Internet radio is as vast as cyberspace itself (for example, Live365 offers more than 30,000 Internet radio broadcasts).

In comparison to traditional radio, Internet radio is **not limited to audio**. An Internet radio broadcast can be accompanied by photos or graphics, text and links, as well as interactivity, such as message boards and chat rooms. This advancement allows a listener to do more than listen. In the example at the beginning of this article, a listener who hears an ad for a computer printer ordered that printer through a link on the Internet radio broadcast Web site. The relationship between advertisers and consumers becomes more interactive and intimate on Internet radio broadcasts. This expanded media capability could also be used in other ways. For example, with Internet radio, you could conduct training or education and provide links to documents and payment options. You could also have interactivity with the trainer or educator and other information on the Internet radio broadcast site.

Internet radio programming offers a **wide spectrum of broadcast genres**, particularly in music. Broadcast radio is increasingly controlled by smaller numbers of media conglomerates (such as Cox, Jefferson-Pilot and Bonneville). In some ways, this has led to more mainstreaming of the programming on broadcast radio, as stations often try to reach the largest possible audience in order to charge the highest possible rates to advertisers. Internet radio, on the other hand, offers the opportunity to expand the types of available programming. **The cost of getting "on the air" is less** for an Internet broadcaster (see the next section, "Creating an Internet Radio Station"), and Internet radio can **appeal to "micro-communities" of listeners focused on special music or interests**.

Creating an Internet Radio Station

Internet Radio is not like traditional radio where you need expensive equipment to broadcast. Today, it is much easier for just about anybody with the right skills to operate their **own Internet radio station**. Unlike broadcast stations, the Internet variety do not require powerful antennas to transmit radio signals. Instead, you operate over the same Internet lines that you use to surf the web.

What do you need to set up an Internet radio station?

• Microphone and headphone – The microphone does not have to very expensive depending on much you plan on using it. If you want to keep your hands free and to be able to move around while talking, you should consider a headset, i.e., headphones with an attached mic. Keep the microphone close to the mouth to reduce the amount of background noise that it picks up.

• Mixer – A mixer allows you to connect multiple microphones and telephone interfaces to your broadcasting system. All your audio inputs go into the mixer and come out as one combined signal. You may want to find a mixer that accommodates USB or other HD digital cables.

• Headphone amplifier – If you will be having multiple hosts and/or guests on your broadcasts, you should have a headphone amplifier. With this equipment, you can give each person their own headphone with with boosted output for each signal. Mixers often have only a single headphone jack, so you will need the amplifier to boost the signal for each pair of headphones.

• Server – A server is simply a computer that you will dedicate to your Internet radio broadcast. While you could use your regular personal computer, if you plan to own a radio station, you will need a dedicated server to provide acceptable quality. The computer should have ample RAM memory, processing power and similar features because serving a radio signal can consume significant hardware resources.

• Internet connection – Your regular Internet connection probably will not suffice for your radio station. Think of having at least a 56 Kbps stream to provide your listeners with good audio quality.

• Computer connection – If your mixer does not have a computer-compatible interface, then you will need a USB or other interface that will allow you to connect the two devices so that you can pump the audio from the mixer into your computer.

• Telephone interface – A telephone interface allows you to take calls from your audience. Obviously, this is vital for Internet radio talk shows. For music stations, it is optional, but many of your competitors will probably be accepting calls. If you only plan to take a call at a time, then you can use a single-line hybrid. To deal with multiple callers on hold, then you will need a talk show system with multiple lines. These systems may require that you install separate analog phone lines at your radio station location. Make sure that your mixer can properly handle "mix-minus" signals if you are using analog phone lines.

• Internet radio apps – The simplest way to stream music is to use a web streaming service like Ustream.tv or Justin.tv. These services allow you to send your audio input to the service's web servers in MP3 format. When you are done live broadcasting, you have the option of hooking up your YouTube playlists to provide listeners with content on a 24/7 basis. For something more sophisticated, you will need apps that allow you to choose and organize music files quickly and that will convert your audio feed into streaming format. Winamp is an example of a music-playing app while Edcast is an audio conversion app. You can use such a setup to create a live stream to services like Icecast. Shoutcast offers a free plugin that you can download to convert your audio into a format for their Internet radio broadcast.

Tip Radio Broadcasting Software- I tried several ways to broadcast our commercial Internet Radio station where we could insert commercials, Drops, shows and more. To be able to automate the station in a cost effective manner. One of the best I used that gave us the tools needed to broadcast with ease and was able to do live remotes is a software call Sam Broadcaster. It's only available on Windows but it's powerful. You can develop playlists, commercials, drops and you can preset your day shows and not have to be there. It will automatically insert commercials, drops just like a real station.

Getting audio over the Internet is pretty simple:

1. The audio enters the Internet broadcaster's encoding computer through a sound card.
2. The encoder system translates the audio from the sound card into streaming format. The encoder samples the incoming audio and compresses the information so it can be sent over the Internet.
3. The compressed audio is sent to the server, which has a high bandwidth connection to the Internet.
4. The server sends the audio data stream over the Internet to the player software or plug-in on the listener's computer. The plug-in translates the audio data stream from the server and translates it into the sound heard by the listener.

There are two ways to deliver audio over the Internet: downloads or streaming media. In **downloads**, an audio file is stored on the user's computer. Compressed formats like MP3 are the most popular form of audio downloads, but any type of audio file can be delivered through a Web or FTP site. **Streaming audio** is not stored, but only played. It is a continuous broadcast that works through three software packages: the encoder, the server and the player. The **encoder** converts audio content into a streaming format, the **server** makes it available over the Internet and the **player** retrieves the content. For a live broadcast, the encoder and streamer work together in real-time. An audio feed runs to the sound card of a computer running the encoder software at the broadcast location and the stream is uploaded to the streaming server. Since that requires a large amount of computing resources, the streaming server must be a dedicated server.

Radio Podcast & How it Works

Podcasts are digital media files (most often audio, but they can be video as well), which are produced in a series. You can subscribe to a series of files, or podcast, by using a piece of software called a podcatcher. Once you subscribe, your podcatcher periodically checks to see if any new files have been published, and if so, automatically downloads them onto your computer or portable music player for you to listen to or watch, whenever you wish.

How Podcasting Works

Have you ever dreamed of having your own radio show? Are you a recording artist hoping to have your songs heard by the masses? Decades ago, you would have needed a lot of connections -- or a fortune -- to get heard.

But now, thanks to the Internet and its instantaneous connection to millions of people, your dreams can become reality. Just as blogging has enabled almost anyone with a computer to become a bona fide reporter, podcasting allows virtually anyone with a computer to become a radio disc jockey, talk show host or recording artist.

Although podcasting first found popularity within the techie set, it has since caught on with the general public. Log on to one of several podcast sites on the Web, and you can download content ranging from music to philosophy to sports. Podcasting combines the freedom of blogging with **digital audio technology** to create an almost endless supply of content. Some say this new technology is democratizing the once corporate-run world of radio.

Podcasting is a free service that allows Internet users to pull audio files (typically MP3s) from a podcasting Web site to listen to on their computers or personal digital audio players. The term comes from a combination of the words **iPod** (a personal digital audio player made by Apple) and **broadcasting**. Even though the term is derived from the iPod, you don't need an iPod to listen to a podcast. You can use virtually any portable media player or your computer.

Unlike Internet radio, users don't have to 'tune in' to a particular broadcast. Instead, they **download** the podcast on demand or subscribe via an RSS (Really Simple Syndication) feed, which automatically downloads the podcast to their computers. The technology is similar to that used by TiVo, a personal video recorder that lets users set which programs they'd like to record and then automatically records those programs for later viewing.

In this article, you'll learn how podcasting works, discover where to find podcasts and how to listen to them. You'll also find out what tools you need to create your own podcast and how to promote it, as well as hear what industry analysts have to say about the future of this burgeoning technology and its counterpart, video podcasting.

Podcasting History

Podcasting was developed in 2004 by former MTV video jockey **Adam Curry** and software developer **Dave Winer**. Curry wrote a program, called **iPodder**, that enabled him to automatically download Internet radio broadcasts to his iPod. Several developers improved upon his idea, and podcasting was officially born. Curry now hosts a show called The Daily Source Code, one of the most popular podcasts on the Internet.

Right now, podcasting is free from government regulation. Podcasters don't need to buy a license to broadcast their programming, as radio stations do, and they don't need to conform to the Federal Communication Commission's (FCC) broadcast decency regulations. That means anything goes -- from four-letter words to sexually explicit content. Copyright law does apply to podcasting, though. Podcasters can copyright or license their work -- Creative Commons is just one online resource for copyrights and licenses.

Although several corporations and big broadcast companies have ventured into the medium, many podcasters are amateurs broadcasting from home studios. Because podcasters don't rely on ratings as radio broadcasters do, the subject matter of podcasts can range from the refined to the silly to the excruciatingly mundane. Podcasters typically cater to a niche group of listeners. By podcasting consistently on one subject, podcasters not only assert their expertise on the subject matter but also draw a loyal and devoted group of listeners.

Consider two popular podcasts: Keith and The Girl is a say-anything-about-anything podcast run by Keith and his girlfriend, Chemda. The podcast's official website touts its expressive (and explicit) freedom, proudly proclaiming "Not held back by the FCC or anyone else." On the other end of the spectrum is A Mysterious Universe, a podcast devoted to examination of the paranormal, UFOs and cryptozoology.

Podcasts are also used for informational and educational purposes -- self-guided walking tours, talk shows and training are all available through podcasting, according to Podcasting Tools.

Several companies are trying to turn podcasting into a profitable business. Podcasting aggregators such as PodcastAlley.com and Podcast.net now feature advertisements. The Podcast Network, based in Australia, runs commercials and sponsorships during its audio broadcasts. Television networks have gotten into the action, too. National Public Radio, the Canadian Broadcasting Corporation and the BBC have begun podcasting some of their shows. Corporations such as Heineken and General Motors have created their own podcasts to attract consumers.

How Do People Listen to Podcasts?

It is very easy to listen to a podcast. Once you master a few simple steps and search techniques, there are virtually no limits to what you can hear.

Tip The largest podcast service is Apple and when a person subscribes to a podcaster it is automatically added to their ITunes account. Subscribers will be able to listen at their leisure.

To listen to a podcast:

1. Go to a podcasting site.
   - Apple
   - Feedburner
   - SoundCloud
   - Amazon S3
   - Podomatic
   - OurMedia
   - Libsyn
   - Podbean
   - Buzzsprout
   - And many more…

2. Click on the hyperlink for each podcast you want. You can listen right away on your computer (Windows, Mac and Linux support podcasting) or download the podcast to your portable media player.

3. You can also subscribe to one or more RSS feeds. Your podcasting software will check the RSS feeds regularly and automatically pull content that matches your playlist. When you dock your portable media player to your computer, it automatically updates with the latest content.

Since the advent of MP3 technology, other audio file types have been created that support different sizes and capacities of streaming audio. These include AAC (Advanced Audio Coding) and WMA (Windows Media Audio). No matter the format of your audio, there is free technology available to make listening possible, such as Windows Media Player, Apple Quicktime, VLC media player or Winamp.

Now that you know how to listen podcasts, the next logical step is to learn where to find podcasts that pertain to your interests. The iTunes Store is one purveyor of podcasts. Also try consulting directories like The Podcast Network or The Podcast Directory, and you are likely to find a listing that intrigues you.

You'll probably find any variety of podcast you want on the Web, but if you can't seem to find what you're looking for, you can create your own podcast with relative ease. Virtually anyone with a computer and recording capabilities can create his or her own podcast. Podcasts may include music, comedy, sports and philosophy -- even people's rants and raves.

Creating Podcasts

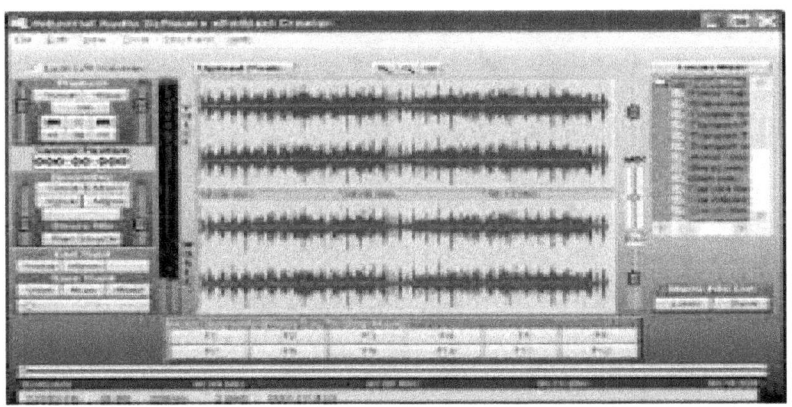

Recording a podcast is almost as easy as listening to one. Here's how the process works:

1. Plug a microphone into your computer
2. Install an audio recorder for Windows, Mac or Linux (free software for audio recorders includes Audacity, Record for All and Easy Recorder V5).
3. Create an audio file by making a recording (you can talk, sing or record music) and saving it to your computer.
4. Finally, upload the audio file to one of the podcasting sites (FeedForAll has a tutorial on how to upload a file).

After recording your podcast, you will want to promote it. FeedforAll and Self Seo offer advice to guide you through this endeavor. Their recommendations include informing the public on how to access and listen to your podcast, advertising your podcast's availability with an eye-catching graphic, writing a press release to notify the public of your podcast and creating a niche group of devoted listeners and assessing their responses to your podcast through installation of a **feedburner link**. A feedburner link keeps tabs on the number of times your podcast has been linked to and any new comments that your listeners have made. This link can be set up through Feed Burner. Tailoring your podcast to your listeners' feedback makes for happy listeners who will, in turn, do the heavy lifting of your promotional work for you.

Several companies are creating new gadgets to improve users' podcasting experiences. San Francisco-based Odeo offers a service that creates custom playlists of podcast files, which can then be downloaded onto portable audio players. Griffin Technology makes a device called Radio Shark 2, which sells for about $50 and can be programmed to record music and radio programs.

Some experts say podcasting still has a long way to go before it catches on with the masses, but its growing popularity is undeniable. It's possible that podcasting will eventually become as popular as text blogging, which grew from a few thousand blogs in the late '90s to more than 7 million today. Some podcasts are already providing thousand of downloads a day, and they're not just entertaining their listeners -- they're also doing business. We'll see how companies are creating and using video podcasts in the next section.

Podcast Tools and Software

A podcast is a non-streamed online digital media file that is delivered by a web-feed also known as an RSS feed. RSS stands for Really Simple Syndication. Many podcasts are delivered via web syndication in installments or episodes often at a given time and place. Podcasts have a number of uses such as training, entertainment, politics, marketing, interviews, storytelling, information dissemination, and many others.

There are many useful tools and software available to use with podcasting. The podcast software and tools can be used to create podcasts, edit podcasts, and deliver podcasts to end users. The software can be broken down into two types of podcasting software - catcher which receives the podcasts and RSS feed, and producer, which can create and edit the podcasts. Here is a list of some of the best podcasting software and tools:

- FeedForAll - Create, edit and publish podcasts.

- AudioBlogger - Use any phone to automatically post audio to your blog.

- Replay Radio - Turn radio broadcasts into podcasts.

- Podcast Producer 2 Podcast Producer 2 is the ideal software for creating, publishing, and distributing podcasts on the Apple platform. It offers a wealth of features to create high-quality podcasts such as picture-in-picture podcasts, dual source capture, and remote control capture via web browsers. Other features include Podcast Composer which has a graphical interface with a variety of editing features and special effects.

- Audacity- Audacity is a great open-source audio editing and recording software program that can be used to create high-quality podcasts. It is available for the Windows, Mac OS X, Linux, and BSD platforms. Audacity has a wealth of editing features that allows amplitude envelope editing, multi-track mixing, unlimited undo of copy, cut, and paste, and tempo editing. It is a great tool for post-processing podcasts.

- ePodcast Producer- ePodcast Producer is a professional podcasting software program with a robust set of integrated podcasting features. In addition to the basic features such as record, edit, and upload, it has an on-screen teleprompter, DSP voice effects, can record telephone interviews with VoIP/Skype recording, and create iTunes tags and RSS feeds. Another great feature of the program is the ability to assign up to 36 music and sound effects to keyboard function keys and on-screen keys.
- ePodcast Creator- ePodcast Creator is a podcast software program that offers a full suite of features that allow you to record, edit, publish, and upload podcasts. It features drag and drop, full iTunes tagging support, voiceover editing, and other features that make this a great tool to develop and deliver podcasts.

- Propaganda- Propaganda is a full-featured and versatile program designed to create, assemble, and distribute podcasts. It offers a great interface that is very user-friendly and geared toward first-time podcast creators. The features include a straightforward publish feature that generates XML, HTML, and MP3 files that are download-ready for podcasts. Propaganda provides all the necessary tools and effects to create a podcast from start to finish.

- Podcaster- Podcaster is a Macintosh based software program that provides a number of features for producing top-notch podcasts. Podcaster can create enhanced podcasts that allow chapter markers to be inserted so that titles, images, weblinks, and other effects can be inserted. The enhanced podcasts allow movies, IPhoto slideshows, and even iMovie projects to be added to the podcast.

- Sennheiser PC 150- A great tool to use for podcasting is the Sennheiser PC 150 headphones. This headset and mic is lightweight and has a comfortable headband, inline volume control, and noise canceling microphone. This is the ideal headset to use along with podcasting software to create high-quality podcasts.

- PodProducer- PodProducer is a very simple podcast producing program that allows the user to edit the audio and add various effects such as cross fade, voiceovers, and other effects. For users looking for podcast software that doesn't have a lot of unnecessary bells and whistles that are found in other podcast software but still gets the job done, then PodProducer is the program to use.

- Winamp- Winamp is a media player designed for the Windows operating system. While is primary features are geared toward playing media files, it can also be used as a podcatcher. Its podcatcher features include using it as an aggregator for RSS media feeds and downloading streaming media.

Conclusion

There are many ways to get your message out but one of the most effective ways to do so. The biggest thing that I want to convey to you as a future Internet Broadcaster is not to be overwhelmed with all the information that is in this guide. It is a reference and a guide full of information for depending on how in depth you want to go. My goal is to give you more so that as you mature and get going you can go to the next level in your journey.

An old friend of mine would say K.I.S.S. (KEEP IT SIMPLE STUPID), and that's where you should start. There is a lot of free software that was in the guide, so start there. If you need help in getting up and running, we offer service aiding those that don't want to do it themselves and we can do it for you and train you on how to work your broadcast and station.

We are also looking for shows, and if you are interested in being a part of our network go to PhatX.net.

To help you set up your own broadcast or show go to www.Briancochran.org or call me at 657-204-6249 for a free consultation.